BREATHE DEEP

AN ILLUSTRATED GUIDE TO THE ≋TRANSFORMATIVE≋ POWER OF BREATHING

MISHA MAYNERICK BLAISE

ADAMS MEDIA
New York London Toronto Sydney New Delhi

Aadamsmedia

Adams Media
An Imprint of Simon & Schuster, Inc.
100 Technology Center Drive
Stoughton, Massachusetts 02072

First Adams Media hardcover edition
May 2023

ADAMS MEDIA and colophon are
trademarks of Simon & Schuster.

For information about special discounts for
bulk purchases, please contact Simon &
Schuster Special Sales at 1-866-506-1949
or business@simonandschuster.com.

The Simon & Schuster Speakers Bureau
can bring authors to your live event. For
more information or to book an event
contact the Simon & Schuster Speakers
Bureau at 1-866-248-3049 or visit our
website at www.simonspeakers.com.

Interior design, illustrations, and
hand lettering by Misha Maynerick Blaise

Manufactured in China

10 9 8 7 6 5 4 3 2 1

Library of Congress Cataloging-in-
Publication Data has been applied for.

ISBN 978-1-5072-2021-4
ISBN 978-1-5072-2022-1 (ebook)

TO MY MOM AND DAD

WELCOME
to the
≋MIRACLE≋
of your
BREATH!

TAKE A DEEP BREATH

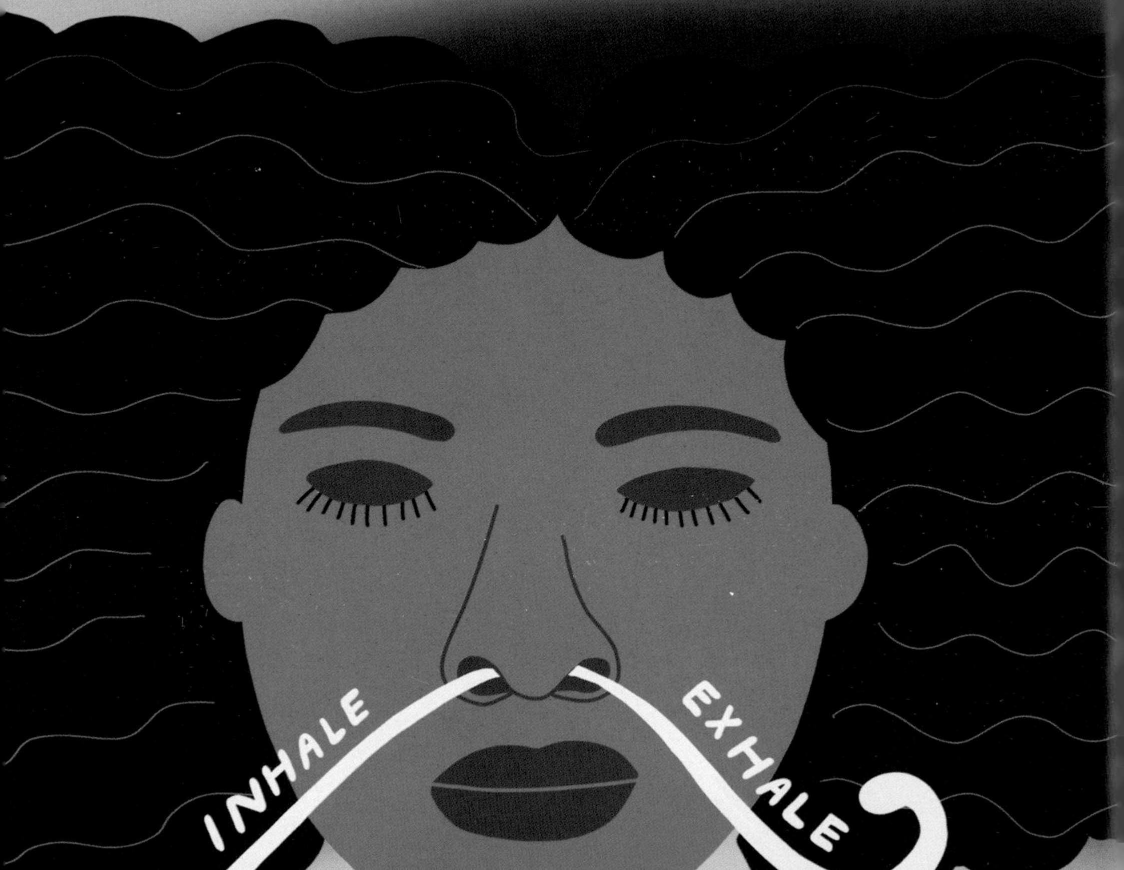

HOW DO YOU FEEL?

Breathing happens.
All day long, awake or asleep, you are sustained by
the movement of air in and out of your lungs.

Inhalation & Exhalation
create a gentle rhythm:

the background soundtrack
to your life.

IF YOU
HAVE
BREATH

YOU HAVE LIFE

The presence of breath is the most vital sign of life.
And if you have life, so much is possible!

Potential things you can do with your life:

GET A CAREER

LEARN HOW TO COOK
SOMETHING BESIDES
SPAGHETTI

ENJOY THE
SIMPLE THINGS

~CONTEMPLATE
THE MYSTERIES~

WEAR PANTS THAT
ARE TOO TIGHT

BE A REBEL!

BECOME
HUGELY
MUSCULAR

CONFORM TO
THE THOUGHTS
AND STYLE OF
THE OTHER REBELS
IN YOUR SOCIAL
GROUP

PAY BILLS... BRUSH TEETH
BE RESPONSIBLE ABOUT STUFF

We are about to embark on information and techniques that will help you unleash

THE SUBLIME POWERS of the BREATH.

BUT FIRST, PLEASE MEET YOUR DIAPHRAGM,

a large, parachute-like muscle that sits just below the lungs and helps facilitate breathing.

Using your abdominal muscles, you can voluntarily control your diaphragm to slow down or speed up your breathing, or to draw in fuller, deeper breaths.

THIS IS HOW IT WORKS!

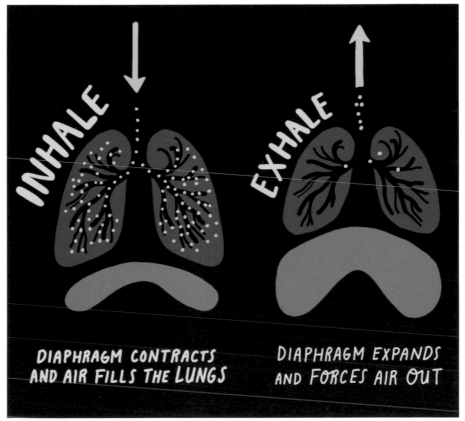

INHALE

EXHALE

DIAPHRAGM CONTRACTS AND AIR FILLS THE LUNGS

DIAPHRAGM EXPANDS AND FORCES AIR OUT

ENGAGING the DIAPHRAGM ALLOWS you to MOBILIZE the TRANSFORMATIVE POTENTIAL of the BREATH.

THE UNIQUE POWER of BREATHING

What's incredible about breathing is that even though it happens automatically, it can also be regulated intentionally. Unlike other automatic functions in the body—like digestion and heart rate—you can consciously choose to control your breath. Amazingly, this, in turn, affects other bodily systems. The heart, brain, nervous system, digestive system, and immune system all have been scientifically proven to be impacted by controlled breathing.

BUNNIES CANNOT CONSCIOUSLY CONTROL THEIR BREATH, BUT WE CAN!

For centuries, people around the world
have consciously used breathwork as a tool
to promote physical health and spiritual
connection. Both ancient wisdom and modern
science confirm that breathwork has a
tremendous impact on both body and mind.

WOW—BREATHING IS SUCH A GIFT!

"BREATHING CONTROL
GIVES... STRENGTH,
VITALITY,
INSPIRATION, AND
MAGIC
POWERS."

—ZHUANGZI,
Chinese philosopher known for writing one
of the foundational texts of Taoism

IT'S NOW TIME TO CONJURE SOME MAGIC POWERS!

BREATHWORK TUTORIAL!

Belly Breathing

TRY IT!

(A.K.A. DIAPHRAGMATIC BREATHING)

Lie down or sit with a straight back. As you take a deep breath in, focus on pulling air into the bottom of your lungs, just above your belly button. In order to do this, you need to engage your stomach muscles and your diaphragm.

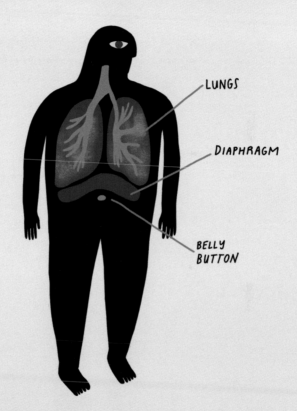

LUNGS

DIAPHRAGM

BELLY BUTTON

The goal is to let your belly expand outward when you inhale, and then pull it inward toward your back when you exhale. If done properly, your stomach will rise and fall with each breath (your chest will barely move).

INHALE: BELLY RISES

2 Hold your breath in for a few seconds.

1 Try inhaling through your nose for a count of four.

EXHALE: BELLY FALLS

3 Exhale for a longer count of eight.

WHILE EXHALING, PURSE YOUR LIPS AND BREATHE OUT A THIN BUT STRONG STREAM OF AIR.

By Consciously Repeating These Deep Belly Breaths For Even Two Minutes, You Activate Your

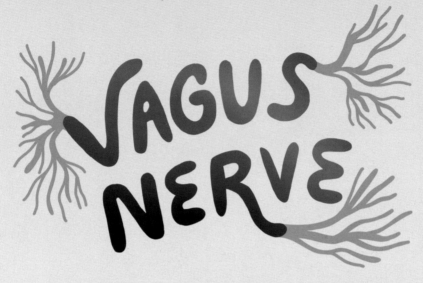

Vagus Nerve

The vagus nerve is a vital part of your nervous system, and the longest nerve that connects the brain to the body. "Vagus" means "wandering" in Latin. The vagus nerve starts at the base of your brain stem and wanders through your face and down across all your major organs until it stops at your colon.

A couple notable roles played by the vagus nerve are that it carries information about pain, touch, and temperature to your brain, and it also helps to slow your heart rate.

The vagus nerve is the main parasympathetic nerve in your body. This means that it stimulates the "rest and digest" mode, also known as "totally chilling out." When the parasympathetic nervous system is activated, your blood pressure lowers, your heart rate slows, digestion is promoted, and you feel blissfully at peace.

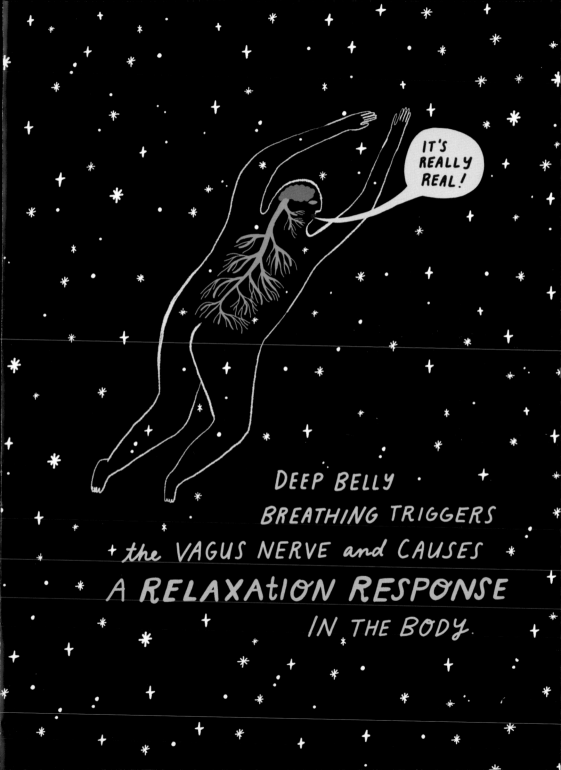

Sudden high-stress situations can, on the other hand, stimulate the sympathetic nervous system, also known as the "fight, flight, or freeze" response. We all know this trash feeling.

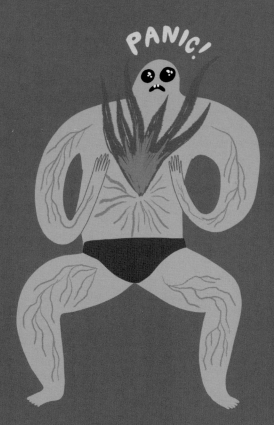

Your nervous system feels like it's exploding, your breathing becomes shallow as your heart rate speeds up, your chest tightens, and you may experience all sorts of unsettling feelings as your brain releases hormones like adrenaline and cortisol.

Obviously, there are times when the stress response to
a threat is totally helpful and even lifesaving.

Stress is a normal part of life, and in good times we can leverage
it toward positive growth. Stress can help release our creative
powers as we rise up to face the demands of life! But chronic high
stress causes all sorts of long-term problems: anxiety, mood swings,
various illnesses, difficulties with digestion and sleep,
and even a doom-and-gloom attitude.

CONTROLLED BREATHING IS AN
Ancient technology
THAT CAN CALM A FRAYED
NERVOUS SYSTEM.

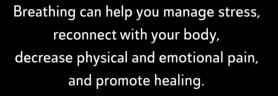

Breathing can help you manage stress,
reconnect with your body,
decrease physical and emotional pain,
and promote healing.

BREATHING CONNECTS BODY, MIND, AND SOUL.

The link between breath and the spiritual dimension of life is rooted in age-old religious texts and wisdom traditions around the world.

IN MANY LANGUAGES, THE WORD FOR "BREATH" IS THE SAME WORD USED TO DESCRIBE *LIFE*, *SOUL*, or *SPIRIT*.

The English word "spirit" is derived from the Latin root word "spir," which means "breathe." The ancient Greek word "pneuma" can mean "breath," "spirit," or "soul."

In the biblical story of Genesis, breath is the vital divine link between humans and their Creator:

"And the LORD God formed man of the dust of the ground, and breathed into his nostrils the breath [neshamah] of life; and man became a living soul."

(GENESIS 2:7)

IN BIBLICAL HEBREW, "NESHAMAH" means both "BREATH" and "SOUL."

.IN SOME CULTURES, BREATH IS SEEN AS A **TOOL** to ACCESS

UNIVERSAL ENERGY.

For thousands of years in both China and India, various spiritual, medical, and martial arts traditions have taught that all of existence is permeated by an eternal vital life force. In Chinese philosophy, this force is referred to as "chi," while in Indian philosophy and yogic practice, it's called "prana." Breathwork is one technique used to harness and draw upon this energy in the human body in order to promote health, longevity, and wisdom.

" PRANA is the BREATH of LIFE of ALL BEINGS IN THE UNIVERSE. THEY ARE BORN THROUGH and LIVE BY It, AND WHEN they DIE their INDIVIDUAL BREATH DISSOLVES INTO THE COSMIC BREATH."

—BKS IYENGAR,
yoga teacher and scholar

ALTERNATE
nostril
BREATHING

TRY IT!

Alternate nostril breathing is a yogic breath control practice. By breathing in a back-and-forth pattern between the nostrils, the mind can focus and calm down. Each nostril corresponds to the opposite hemisphere of the brain; therefore, this technique is thought to balance the energy in the body.

Here is one variation of this practice:

HOLD YOUR RIGHT
HAND LIKE THIS:

Now let the pointer and middle fingers of your right hand rest between your eyebrows.

USE YOUR THUMB TO CLOSE YOUR RIGHT NOSTRIL.

USE YOUR RING FINGER TO CLOSE YOUR LEFT NOSTRIL.

1 Use your thumb to close your right nostril. Inhale slowly through your left nostril for a count of four.

2 Use your thumb and ring finger to close both nostrils so that you can hold your breath for a count of four.

3 Exhale through the right nostril for a count of four, and again close both nostrils and hold for a pause.

4 Now inhale through the right nostril, hold, exhale from the left nostril, and begin the cycle again.

You can increase the length of your inhalations and exhalations—just make sure the pattern is consistent. Try this practice for three to five minutes.

PAY ATTENTION to the EFFECTS on your BODY AND MIND!

THE POWER OF NOSE BREATHING:

In general, breathing through the nose is radically better than breathing through the mouth. The nose is uniquely designed to interface with the outside world so it can bring the best quality air into the lungs. Nasal hairs filter dust, allergens, and pollen.

As you inhale, your nose warms and moisturizes the air to keep your lungs from drying out. Nose breathing also promotes deeper breaths into the lower lungs (stimulating the healing powers of your best friend: the parasympathetic nervous system!).

But probably the most stunning aspect
of nasal breathing is that it naturally
releases a powerful gas called

NITRIC
OXIDE

A.K.A.
"THE MIRACLE MOLECULE"

Nitric oxide has been called the miracle molecule because of its
many positive impacts on the body. It has antiviral and antibacterial
properties—and boosts your immune system. It plays an important
role in dilating the blood vessels in the lungs so that oxygen and
nutrients can effectively travel to every part of the body.

Humming

TRY IT!

Research shows that humming greatly increases the production of nitric oxide in the nose! In fact, humming causes a fifteen- to twenty-fold increase in nitric oxide compared with normal nose breathing.

You for sure already know how to hum, but here is some humming advice: Rest the tip of your tongue just behind the two top front teeth. Don't push the air out forcefully; let it flow out gently. Breathe in through your nose. Enjoy your hum!

We typically think of breathing as a very simple exchange:
We bring in GOOD oxygen, and expel BAD carbon dioxide.

In reality,
CARBON DIOXIDE IS ALSO GOOD!
It plays a vital part in delivering oxygen to our cells.

TOO MUCH!

It sounds counterintuitive, but mouth breathing can cause overbreathing, which brings in too much oxygen, and depletes carbon dioxide.

Taking shallow breaths into the upper lungs only makes things worse. If done all the time, these poor habits can lead to chronic hyperventilation. This contributes to anxiety, weakened immunity, poor sleep, and a whole bunch of other issues!

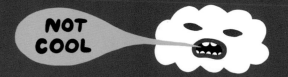

NOT COOL

BREATHING TIPS
FOR EVERYDAY LIFE:

- Rest the tongue on the upper palate of the mouth, just behind your two front teeth.

- Breathe through your nose as much as possible.

- Breathe gently, slowly, and deeply toward the lower part of your lungs.

- Strive for a shorter inhale and a longer exhale.

RELAX
AND
ENJOY

With each breath, oxygen enters your lungs, and diffuses through the blood to every single cell in your body!

(DANG, THAT OXYGEN IS GOING EVERYWHERE!)

P.S.: YOUR BODY HAS OVER THIRTY TRILLION CELLS!!

THE BREATH
TEACHES US

MINDFUL BREATHING PROVIDES AN OPPORTUNITY to Strengthen THE MIND-BODY CONNECTION AND **REWIRE** the **BRAIN**.

One of the most exciting breakthroughs in brain science in the past century is the discovery that the brain has an innate power to change, adapt, and restructure itself in response to its environment.

This is known as

"NEUROPLASTICITY."

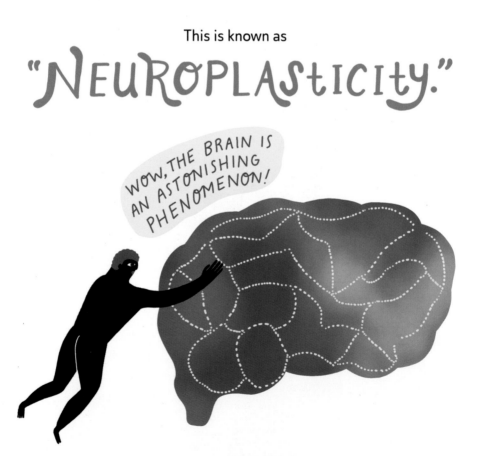

NEUROPLASTICITY IS PRETTY MUCH

THE COOLEST THING EVER!

It's a cosmic superpower that opens infinite possibilities for profound transformation and healing!

A BASIC INTRO TO HOW NEUROPLASTICITY WORKS

I FEEL SCARED

All your thoughts, feelings, habits, and behaviors enforce the circuitry in your brain known as NEURAL PATHWAYS.

Your brain forms these neural pathways the same way hikers impact trails through the forest. Paths rarely used become overgrown, hard to find, and soon disappear. But the more hikers go on a single path, the deeper, wider, and more established that path becomes. Similarly, everything you think, feel, or do impacts the brain's neural pathways.

INCREDIBLY, NEW NEURAL CONNECTIONS ARE FORMED THROUGHOUT YOUR ENTIRE LIFE!

Science shows that you can literally create new neural pathways based on where you choose to focus your attention. New activities and habits of thought create stronger neural pathways over time with repetition. Neurons that fire together wire together. In other words, the brain changes—depending on how you use it!

The problem is, if you are feeling fearful or disconnected, it's not easy to just instantly switch modes of thought and jump onto a better path. Old habits are especially hard to change. This is when the powers of the breath can be of great help.

Mindful breathwork is a tool that can assist you to clear the way for positive neuroplastic transformation, so you can rewire your brain.

THE PHYSIOLOGICAL CHANGES BROUGHT BY INTENTIONAL BREATHWORK CAN

TRANSFORM YOUR STATE OF MIND.

Slow and deep breathing can lower stress chemicals in the body, and turn off the fight, flight, or freeze response. When your nervous system feels safe, the body returns to a state or rest and repair, which makes it much easier to focus on what allows you to feel hopeful, creative, empowered, and grateful.

BREATHWORK TUTORIAL!

BUILDING NEW NEURAL PATHWAYS THROUGH

TRY IT!

Mindful Breathing

(Inspired by Alan Gordon, LCSW, executive director of the Pain Psychology Center in Los Angeles)

Set aside a few minutes, get comfortable, and direct your attention to the positive sensations felt when the breath enters and exits the body.

The most important thing in this exercise is to let the act of breathing communicate a message of safety to your nervous system.

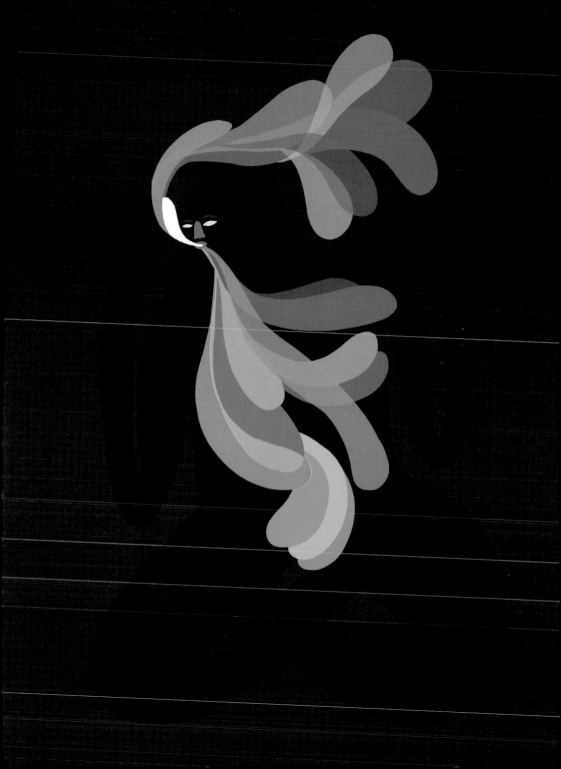

TO PRACTICE MINDFUL BREATHING, YOU SHOULD *Consciously* RECEIVE YOUR BREATH with an OPEN and GENTLE CURIOSITY.

Feel how the breath expands your lungs and belly. Notice how the breath affects your nasal passages. Observe the slight change in temperature: cooler on the inhale, warmer on the exhale.

Enjoy the good feeling of the breath through a lens of safety. Let your breath wrap you in a big, warm bear hug (a cute and safe bear). If your mind wanders, or you find an uncomfortable sensation in your body, that's okay. Just continue to bring your attention back to the nice feeling of your breath.

THAT'S IT! That's all you have to do.

you are safe

This intentional act of relaxed enjoyment communicates messages of trust and safety to your brain. It reinforces neural pathways that calm the nervous system.

You are literally changing the structure of your brain through conscious action! The more you practice this simple technique, the stronger the neural pathway becomes, and the easier it is for your mind to return to the breath to calm down.

"BREATH *is the* BRIDGE
WHICH CONNECTS
LIFE *to*
CONSCIOUSNESS,
WHICH UNITES *your*
BODY *to your* THOUGHTS.
WHENEVER *your* MIND
BECOMES SCATTERED,
USE *your* BREATH
AS *the* MEANS *to* TAKE HOLD
of your MIND AGAIN."

— Thích Nhât Hạnh,
meditation teacher, poet, and peace activist

THAT SOUNDS GREAT!

BONUS EXERCISE!

As you enjoy your breath, comfort yourself with
messages of encouragement:

THE FACT THAT I'M BREATHING
IS PROOF THAT the UNIVERSE
IS StILL SUPPORTING ME.

WHATEVER
HAPPENS,
IT WILL
BE OKAY.

I CAN LET LIFE
POUR THROUGH ME.

I CAN RIDE THE WAVE OF
LIFE'S HIGHS and LOWS.

I CAN USE MY
EXPERIENCES
to LEARN
AND GROW.

I AM OPEN to
RECEIVING ALL THE
GOOD THINGS LIFE
HAS to OFFER.

I APPRECIATE the
MYSTERIOUS GIFT
OF LIFE.

YOU GOT THIS, BABÉ.

YOUR BREATH IS LIKE A
GOOD FRIEND WHO IS
ALWAYS THERE FOR YOU.

YOU CAN COMMUNE with your FRIEND IN SECRET ANYTIME AND ANYWHERE!

WHILE STANDING IN LINE...

DURING A BORING MEETING...

IN THE MIDDLE OF A LARGE CROWD.

THE BREATH IS A
SACRED POWER
THAT CONNECTS
US WITH
ALL of
EXISTENCE.

Right now, you are in intimate communication with the world around you. The air physically touches every part of your body. It moves inside of you and penetrates every cell, simultaneously surrounding your outer being.

THE MOLECULES you BREATHE PARTICIPATE IN GRAND *geological* AND *chemical* PROCESSES; THEY ARE RECYCLED THROUGH the land, water, and atmosphere, MOVING THROUGH THE WORLD over the course of MILLIONS of YEARS.

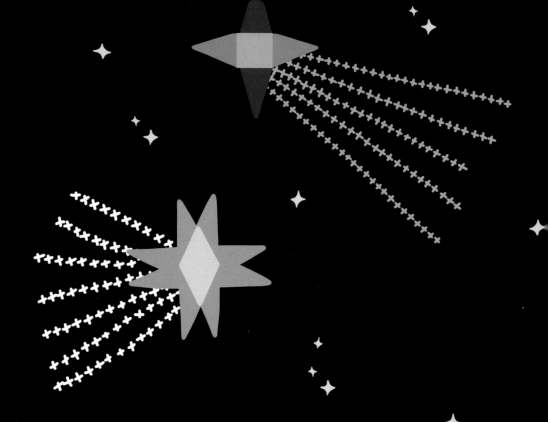

The air we breathe is ancient, originating
from the heart of burning stars, just like us.

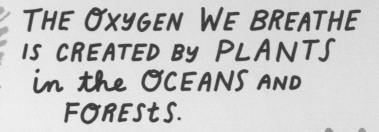

THE OXYGEN WE BREATHE IS CREATED BY PLANTS in the OCEANS AND FORESTS.

At every moment,
our breath brings us into
PROFOUND COMMUNION
with the earth and
ALL OF CREATION.

OUR BREATH CONNECTS US WITH OtHER PEOPLE.

Although it's invisible to the human eye, each person is surrounded by a microbial cloud: an assortment of viruses, bacteria, and fungi that swirl in the air up to a few feet around the body. When you pass by another person, their cloud rains upon your skin and is breathed into your lungs.

WHEN WE SHARE SPACE WITH OTHERS, WE LITERALLY **BREATHE EACH OTHER'S MICROBES** AND EXCHANGE PARTS OF OURSELVES IN THIS **DISTURBINGLY INTIMATE WAY.**

This process of physical interchange with each other
and our shared universe is symbolic of a larger truth:
Humanity is totally interconnected.

WE ARE ONE HUMAN FAMILY SHARING ONE PLANETARY HOMELAND.

INtERPERSONAL SYNCHRONIZATION

is the phenomenon in which individuals begin to physiologically mirror the people they're with. Studies show that quietly sitting next to another person is enough to generate synchronicity in respiratory and heart rates.

COULD OUR OWN CALM NERVOUS SYSTEMS BRING CALM to THOSE AROUND US?

Is breathwork something that impacts not just the individual, but also those we are with?

THE *Physiological Sigh*

TRY IT!

This technique is a simple and powerful way to calm
the body and mind very quickly.

1 Take a strong inhalation through your nose or mouth.
At the top of your inhalation, inhale one more time to take in as
much air as possible.

2 Release your breath in a prolonged exhalation from your mouth.

JUST LET GO...

Just a few rounds of these
breaths will cause an immediate
physiological effect on your body,
which in turn will help bring peace
to your mind. Maybe through
interpersonal synchronization,
your own state of calm can bring
calm to others.

The physiological sigh is a potent practice that calms the nervous system right away. This is because it activates some adorable members of your respiratory system:

THE ALVEOLI.

HI!

Alveoli are balloon-like sacs at the end of the branches in your lungs (technically they are at the end of the bronchioles, which are the tiny branches coming off of the bigger tubes, or bronchi, in the lungs).

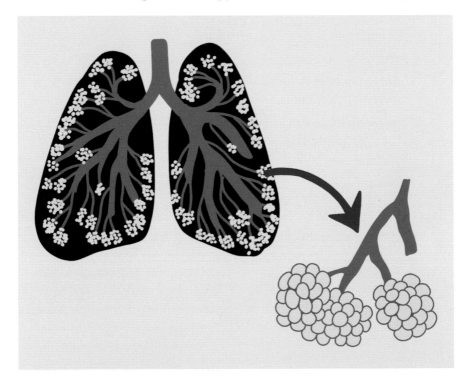

These little guys are the transitional site where oxygen and carbon dioxide molecules move in and out of your bloodstream.

OUR LUNGS CONTAIN 600 MILLION ALVEOLI!

UNFATHOMABLE!

BUNNIES CANNOT COMPREHEND THAT NUMBER, AND NEITHER CAN WE!

If you spread the alveoli all out, they would cover the surface area of an entire tennis court!

These sacs sporadically collapse, which causes a buildup of carbon dioxide in the bloodstream. The body's natural way to reinflate the sacs is through sighing. Believe it or not, you sigh every five minutes or so, day and night, in order to pop the alveoli back open and prevent your lungs from collapsing.

When you are stressed, your breathing becomes shallow, causing excessive carbon dioxide to build up in the body. This causes alveoli to collapse at a higher rate. The buildup also helps trigger a stress response. A deliberate sigh is a dynamic way to quickly pop those alveoli back open, reduce carbon dioxide in your bloodstream, and quiet the nervous system.

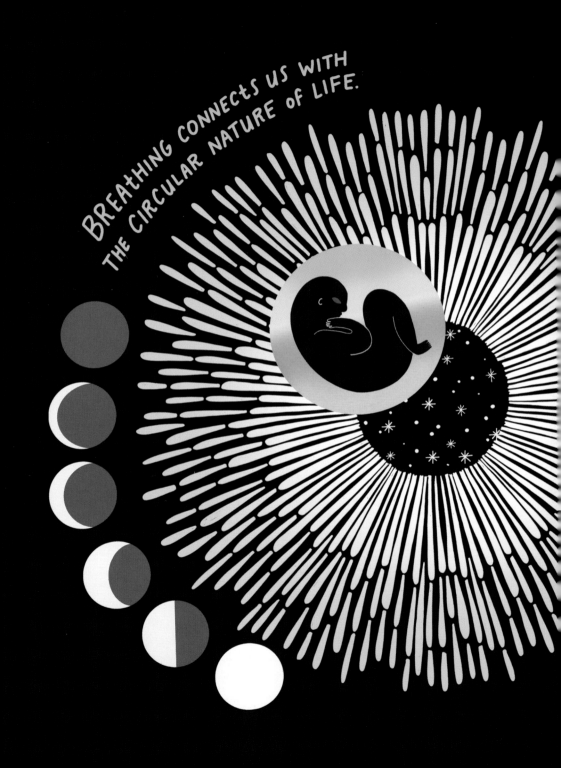

BREATHING CONNECTS US WITH THE CIRCULAR NATURE of LIFE.

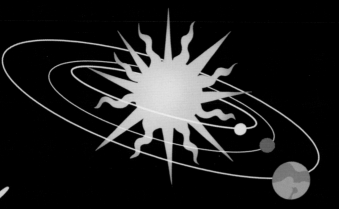

THE BREATH SYMBOLIZES
THE **PRIMAL PATTERN**
OF **EXISTENCE:**

GROWTH (INHALE)
PRESERVATION (BREATH HOLD)
DECAY (EXHALE)
REGENERATION (ANOTHER BREATH!)

LIFE·DEATH·REBIRTH

BREATHING ALSO Puts us IN TOUCH WITH *the* LINEAR NATURE *of* LIFE.

LIFE BEGINS with OUR FIRST BREATH AND WILL END with OUR LAST, AND THERE ARE ONLY SO MANY BREATHS IN BETWEEN.

WITH EVERY BREATH WE TAKE, OUR BODY AGES EVER SO SLIGHTLY IN ITS ONWARD PROGRESSION TOWARD ITS FINAL FATE.

Breath connects us with time itself: With each breath, we take in particles created during the ancient beginnings of the universe. We breathe in molecules that have cycled through other beings and elements of existence across all of history.

OUR BREATH
CONNECTS US
WITH the
PAST and the
FUTURE...

...AND WE BREATHE
IN THE ETERNAL
NOW MOMENT.

BREATHING

LIFE IS HARD,
AND OUR BREATH
CAN HELP US
DO HARD THINGS.

IT HURTS SO MUCH

Uncertainty, difficulty, and pain
are all part of the human condition.

So is the potential for love, connection, healing, and resiliency.

Our breath is there for us through it all, sustaining us during both our highs and lows, allowing us the opportunity to learn and have another chance.

PART of the ART of BEING HUMAN IS LEARNING HOW BEST to RIDE ALONG the EVER-SHIFTING RIVER of LIFE.

LIFE IS ALL ABOUT
CHANGE and TRANSITION.
JUST LIKE OUR BREATH,
WHATEVER WE TAKE IN,
WE WILL HAVE TO
EVENTUALLY
LET GO.

EVERYTHING IS IMPERMANENT.

BREATHWORK HAS
BEEN USED FOR
MILLENNIA TO
HELP PEOPLE

MANEUVER
THROUGH

THE HUMAN
PREDICAMENT.

You already know that breath can be used to help you calm your nervous system so you can more easily experience positive feelings like safety and connection.

BREAthWORK CAN ALSO BE USED STRATEGICALLY to ≣STRESS OUt≣ YoUR NERVOUS SYStEM IN A WAY THAT MAY BETTER PREPARE yoU to HANDLE TOUGH TIMES.

HORMETIC STRESS

is a short burst of stress that someone intentionally chooses to expose themselves to—and that, if done at a higher dose, would be harmful. That sounds scary, but consider a few examples: a two-minute cold shower, a short period of fasting, or a high-intensity interval training (HIIT) workout.

A SHORT BURST OF STRESS!

For someone without high-risk health conditions, these things may be difficult or uncomfortable, but they are not dangerous. They are just enough of a shock to disrupt the homeostasis of the body.

THIS CAN GIVE A QUICK JOLT OF ENERGY AND A SHIFT IN MOOD.

I FEEL ALIVE!

And researchers have found that it can cause a whole range of effects at a cellular level that stimulate the immune system, slow aging, and better prepare you to handle future stress— both physical and mental.

"SHORT-TERM STRESS TRAINING MAY ELEVATE POSITIVE MOOD and MAKE US LESS VULNERABLE to FEELING ANXIOUS OR DEPRESSED WHEN STRESSFUL THINGS HAPPEN IN OUR LIFE."

—ELISSA EPEL, professor of psychiatry at the University of California, San Francisco

THIS IS SOME ANCIENT STUFF

The use of breathwork to stimulate hormetic stress is a very ancient practice, with many techniques alive today—each with different aims and benefits. The pranayama technique "bhastrika" ("breath of fire"), along with the Tibetan Buddhist method of "tummo breathing" (adapted and popularized by Wim Hof), are examples of techniques that promote a burst of high oxygen and low carbon dioxide.

On the opposite end are techniques that lower oxygen and increase carbon dioxide. Some of these exercises are included in the popular Buteyko method.

SIMPLE BREATH HOLD*

TRY IT!

This technique provides a short burst of elevated carbon dioxide.

(Note: This technique should not be used by pregnant women or anyone with serious health conditions. Do not try while swimming, driving, etc.)

1 Take a normal inhale and exhale through the nose.

2 Pinch your nose closed and hold your breath.

3 While holding your breath, rock back and forth or walk around.

4 Continue the breath hold until you feel uncomfortably hungry for air; then release the nose.

5 Breathe again slowly through your nose. Wait at least thirty seconds, then repeat the technique several more times.

6 Return to normal breathing.

BREATHING IS POWER!

*Adapted from Patrick McKeown, teacher of the Buteyko method and director of education and training at Oxygen Advantage

"THE **BREATH** KNOWS HOW TO GO DEEPER **THAN THE** MIND."

– WIM HOF,
motivational speaker
and extreme athlete

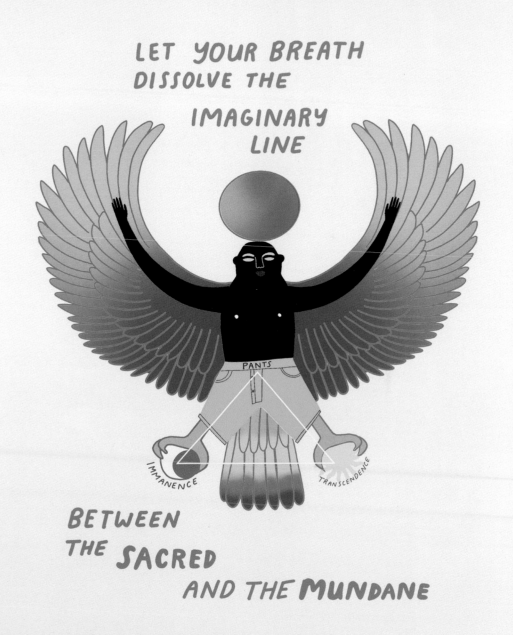

YOU
MATTER.

You have emerged from
a mysterious cosmos,
the result of the perfect
unfolding of events from
the beginning of time. In
this moment, your inhale
and exhale give proof
that the universe is still
supporting your existence.

HOW *do* YOU
WANT to
BETTER YOURSELF?

WHERE *do* YOU WANT to
FOCUS YOUR ATTENTION?

WHAT *do* you WANT
to LEARN MORE ABOUT?

WHAT *do* YOU WANT to DO
WITH YOUR PRECIOUS
LIFE ENERGY?

BREATHE IN

AND AFFIRM YOUR
DESIRED INTENTIONS—

YOUR HOPES
and DREAMS.

BREATHE OUT

and RELEASE your NEGATIVE or LIMITING BELIEFS.

Let your BREATH BE A SOURCE of STRENGTH to CARRY you FORWARD.

THE BREATH
can guide you
SAFELY to
YOURSELF.

WITHIN YOU IS A SAFE PLACE
NO ONE CAN TAKE AWAY,
A PLACE WHERE YOU BELONG AS A
SACRED PART
OF A
VAST
AND
ILLIMIT-
ABLE
CREATION.

A PLACE WHERE YOU ARE
FUNDAMENTALLY OKAY, AND
IT'S ALL GOOD.
Let your Breath
guide you to that place.

THE BREATH is a PORtAL to REALMS of *Mystical* CONNECtION.

In religious and spiritual traditions the world over, the breath is used to connect the individual with the Divine. Whether through prayer, chant, song, or silent contemplation, physical breath represents our utter dependence on some greater source.
What higher power is breathing into us?
How do we each worship and give thanks for the gift of life?

THE BREATH CAN HELP
us IMAGINE
AND CREATE
A NEW WORLD TOGETHER.
WE ARE ALIVE AND BREATHING
IN A TIME OF HUMAN HISTORY
UNLIKE ANY OTHER.
IN THIS ERA,
WE HAVE CLEAR PROOF
OF THE ONENESS
OF HUMANITY.

"As nations and individuals, we are interdependent....
We are all caught in an inescapable network of mutuality,
tied into a single garment of destiny. Whatever affects
one directly, affects all indirectly....We aren't going to have
peace on Earth until we recognize this basic fact of the
interrelated structure of all reality."

—MARTIN LUTHER KING JR.,
minister and civil rights activist

WE EXIST IN RELATION
to ONE ANOTHER,
SO it MAKES SENSE THAT
BUILDING A MORE PEACEFUL
SOCIETY IS
INTERRELATIONAL.

WE ARE ALL IN
THIS TOGETHER.

When the world seems overwhelmed by the forces of disintegration, it can be hard to imagine creating something different. But neuroscience shows we can consciously direct our energy toward positive transformation and renewal.

Conscious breathwork is a tool to access the nervous system and our innate powers to be flexible, change course, and adapt. The breath opens space for the creativity, imagination, and hope that's needed to heal ourselves and our communities.

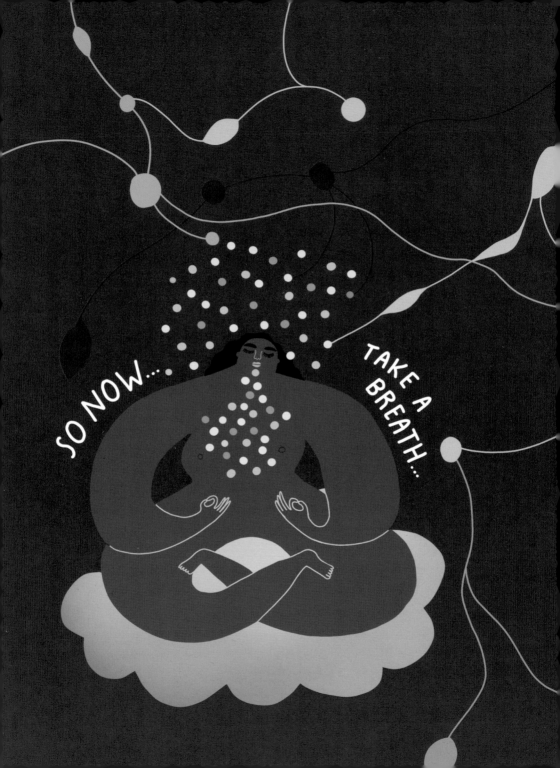

INHALE

AND ACKNOWLEDGE
THE SACRED WEB OF

CONNECTION

THAT UNITES YOU WITH
ALL LIFE ON EARTH.

EXHALE

AND BESTOW COMPASSION
UPON YOURSELF and ALL PEOPLE.

You exist here now, alive in an
era of rapid change, breathing
alongside billions of others
within our shared atmosphere.

OOOOOMM

CHANT "OM,"
THE PRIMORDIAL SOUND
OF THE UNIVERSE,

AND INVOKE THE INHERENT ONENESS
OF ALL THINGS.

MAY HUMANITY'S MATERIAL
ADVANCEMENT BE BALANCED
BY A SPIRITUAL AWAKENING
THAT ALLOWS US TO
TRANSCEND the LIMITING PREJUDICES
THAT DIVIDE US as a HUMAN RACE.

MAY we CONTINUE to LEARN to LIVE
BY the PRINCIPLES of
RECIPROCITY, JUSTICE, GENEROSITY,
and UNITY.

MAY WE BE GUIDED BY
THE SOUL of LIFE
to EXPERIENCE AND SHARE
the LOVE
WE WERE CREATED FOR.

"LOVE IS THE BREATH OF THE HOLY SPIRIT IN THE HEART OF MAN."

-THE BAHÁ'Í WRITINGS

ENDNOTES

Threat and the Autonomic Nervous System

The autonomic nervous system (ANS) manages two separate systems that help us perceive and respond to cues of threat or safety:

1. **The sympathetic nervous system:** Responsible for arousing and mobilizing the body in times of high activity or threat. This system prompts the fight, flight, freeze, or fawn response.
2. **The parasympathetic nervous system:** Responsible for calming our bodies and conserving energy. As the body slows down, it enters a mode of rest and repair, also known as rest and digest.

As we go through the day, we constantly scan the environment for signs of safety or threat. The sympathetic nervous system reacts involuntarily to perceived threat. However, things like past trauma, phobias, a hypersensitive temperament, and psychological disorders can oversensitize some people to perceived threat. Breathwork is one tool that can be used to calm the ANS and activate the parasympathetic nervous system.

The Fight, Flight, or Freeze Response

This describes an immediate physiological reaction to perceived threat. It is also commonly referred to as fight, flight, freeze, or fawn. These reactions are basically understood as follows:

- **Fight:** Attacking any perceived threat aggressively.
- **Flight:** Running away from the threat.
- **Freeze:** Being immobilized in the presence of a threat.
- **Fawn:** Attempting to please and appease the threat and thus avoid any conflict.

Alternate Nostril Breathing

A collection of research regarding this technique can be found at the *National Library of Medicine* site:
https://pubmed.ncbi.nlm.nih.gov/?term=alternate+nostril+breathing

The Power of Humming

Weitzberg, E., & Lundberg, J.O. (2002). "Humming Greatly Increases Nasal Nitric Oxide." *American Journal of Respiratory and Critical Care Medicine, 166*(2), 144–145. https://doi.org/10.1164/rccm.200202-138BC

Overbreathing

Chenivesse, C.; Similowski, T.; Bautin, N.; Fournier, C.; Robin, S.; Wallaert, B.; & Perez, T. (2014). "Severely Impaired Health–Related Quality of Life in Chronic Hyperventilation Patients: Exploratory Data." *Respiratory Medicine, 108*(3), 517–523. https://pubmed.ncbi.nlm.nih.gov

McKeown, P., & Macaluso, M. (2017, March 9). "Mouth Breathing: Physical, Mental, and Emotional Consequences." *Oral Health Group.* www.oralhealthgroup.com/features/mouth-breathing-physical-mental-emotional-consequences/

Building New Neural Pathways Through Mindful Breathing

This breathwork tutorial was inspired by a guided breathing exercise by Alan Gordon, LCSW, called "Leaning Into Positive Sensations."
I encountered it on the Curable app (www.curablehealth.com), which uses modern neuroscience research in the treatment of chronic pain. I would like to note that focusing on positive sensations is just one aspect of a much larger approach to dealing with chronic pain and anxiety. Equally

important are techniques proven to help confront and process unpleasant emotions like anger, fear, and sadness that may underlie chronic pain, anxiety, and other symptoms. In addition to the Curable app and Alan Gordon's book *The Way Out*, I also highly recommend the work of Nicole Sachs, LCSW (thecureforchronicpain.com), Les Aria, PhD (https://lesariaphd.com), and David Hanscom, MD (https://backincontrol.com).

Interpersonal Synchronization

Goldstein, P.; Weissman-Fogel, I.; & Shamay-Tsoory, S.G. (2017). "The Role of Touch in Regulating Inter-Partner Physiological Coupling During Empathy for Pain." *Scientific Reports, 7*(1), 3252. https://doi.org/10.1038/s41598-017-03627-7

The Science Behind Sighing

Some new science related to sighing is based on research from neurobiologist Jack Feldman's lab at UCLA and biochemist Mark Krasnow's lab at Stanford.

Li, P.; Janczewski, W.A.; Yackle, K.; Kam, K.; Pagliardini, S.; Krasnow, M.A.; & Feldman, J.L. (2016). "The Peptidergic Control Circuit for Sighing." *Nature, 530*(7590), 293–297. https://doi.org/10.1038/nature16964

The Physiological Sigh and Stress Reduction

Huberman, Andrew. (2022, April 7). "Reduce Anxiety & Stress with the Physiological Sigh | Huberman Lab Quantal Clip" [Video]. *YouTube*. https://youtu.be/rBdhqBGqiMc

Hormetic Stress

Bai, N. (2019, March 22). "Can Short-Term Stress Make the Body and Mind More Resilient? A New Study Is Testing That Theory." *UCSF*. www.ucsf.edu/news/2019/03/413711/can-short-term-stress-make-body-and-mind-more-resilient-new-study-testing-theory

Gems, D., & Partridge, L. (2008). "Stress-Response Hormesis and Aging: 'That Which Does Not Kill Us Makes Us Stronger.'" *Cell Metabolism, 7*(3), 200–203. https://doi.org/10.1016/j.cmet.2008.01.001

Mattson M.P. (2008). "Awareness of Hormesis Will Enhance Future Research in Basic and Applied Neuroscience." *Critical Reviews in Toxicology*, *38*(7), 633–639. https://doi.org/10.1080/10408440802026406

Breath Holding

The breath hold technique I featured in this book is adapted from a tutorial given by Patrick McKeown; it can also be used to clear up a stuffy nose. See:

Oxygen Advantage. (2017, December 18) "Nose Unblocking Exercises—How to Get Rid of a Blocked Nose" [Video]. *YouTube*. https://youtu.be/cU4ls5ku4Rg

McKeown offers many techniques to correct overbreathing in his book *The Breathing Cure: Develop New Habits for a Healthier, Happier, & Longer Life* (Humanix Books, 2021). See more about the Buteyko method at www.consciousbreathing.com or https://buteykoclinic.com.

See also:

Navarrete-Opazo, A., & Mitchell, G.S. (2014). "Therapeutic Potential of Intermittent Hypoxia: A Matter of Dose." *American Journal of Physiology: Regulatory, Integrative and Comparative Physiology*, *307*(10), R1181–R1197. https://doi.org/10.1152/ajpregu.00208.2014

Bahá'í Quote

This quote is from a talk given by the great teacher 'Abdu'l-Bahá 'Abbás during his visit to Paris in 1911 after forty years of incarceration in the city of Akka. It was recorded in a book called *Paris Talks* (Bahá'í Publishing, 2011).

Finally, I would like to shout out an excellent book describing one man's journey to learn everything about the power of breathing: *Breath* by James Nestor (Riverhead Books, an imprint of Penguin Random House LLC, 2020).

THANK YOU

I would like to express my appreciation to the following people who helped bring this book into reality:

To my agent, Jennifer Weltz, who has diligently worked with me on countless different ideas and proposals over the years! Thanks for your careful attention and dedication, and for always helping to bring out the best in my work.

To my editor, Brendan O'Neill, for believing in this project and being so in tune with my artistic vision.

To the design team at Adams Media: Frank Rivera for his thoughtful feedback, and Priscilla Yuen for her incredible sense of color, composition, and detail.

To my book coach, Marni Seneker, for being a magician who merges intuition with practical guidance. Thank you for helping me bring my ideas to life.

The idea for this book emerged because of my own family's journey into the effects of disordered breathing. For many years, my son suffered from unexplained respiratory problems and mysterious illnesses. We saw many different health professionals who could not diagnose his core problem, yet recommended various drugs and surgeries to handle his symptoms. I would like to thank Dr. Darius Loghmanee, who was the first medical professional to ask the simple question: "Does your son sleep with his mouth open?" From this point, we sought help from myofunctional therapists, who specialize in retraining the muscles of the face and mouth to facilitate healthy breathing. Thanks to Tracey Brizendine in Austin, Texas, and Le Donna Enneking in Fayetteville, Arkansas, who have both helped us to understand the impact of proper breathing on childhood health and development.

To the Moghaddam family for your love, support, and friendship as we have made a life in northwest Arkansas. You are all a blessing, and bring so much joy to our life.

To Donesh Ferdowsi for reading through my rough drafts and being a sounding board for all my ideas and manias. You're a mystical guide and a true friend.

To the friends I've been in deep conversation with this past year: Leila, Emma, Mackenzie, Derik, Narmin, and Felice; thanks for your undying support.

To my kids, Zarek and Kazimir, for your humor, sweetness, and unexpected dance moves that keep me going every day. Finally, to my husband, Nick, for your generous spirit and loving heart. I love our life together.

ABOUT THE AUTHOR

Misha Maynerick Blaise is the author-illustrator of several books including *This Book Is Made of Clouds* and *Crazy for Birds*. Her book *This Phenomenal Life* has been translated into six languages and was a bestseller in China. Misha served as a jury member for the 2017 Golden Pinwheel Young Illustrators Competition at the Shanghai International Children's Book Fair. She lives with her husband and two sons in northwest Arkansas. Visit her at MishaBlaise.com.